Bright White Light II

Written By

Joseph C Jeluso

Contents

About the Author

After devoting a significant portion of my life to specialized professions and earning a college degree, I discovered a passion for writing that led me to where I am today.

After publishing my first novel, “The Bright White Light,” I can share my love for literature with readers worldwide.

I’m also fortunate to share my voyage with an exceptional companion, with whom I embrace life’s excursions and seek out new growth opportunities.

Chapter 1: Honeycomb

I recently stumbled upon a game-changing revelation that has completely transformed my cornbread experience. Allow me to paint the scene for you: I had just returned home, exhausted from a long day, and in my hands, I held a tantalizing takeout dinner from our beloved local restaurant. Their culinary creations never failed to delight, but alas, there was one consistent disappointment—the cornbread. Once again, it arrived dry and lackluster.

Imagine a rustic golden piece of cornbread, resembling the arid sands of a desert, crumbling at the gentlest touch. Its taste was pleasant, but it felt like a promise unfulfilled, missing that certain something. It was in this moment of culinary disappointment that I felt compelled to take matters into my own hands.

With renewed determination, I reached into the cupboard, my fingers wrapping around the honey jar. The amber liquid glowed with warmth, promising to bring a new dimension to the dry cornbread. Taking charge of the situation, I drizzled the honey generously over the cornbread, watching with anticipation as it cascaded and seeped into every nook and cranny, saturating the previously parched surface.

As the honey infused the cornbread, a magical transformation unfolded before my eyes. What was once dry and crumbly, akin to sun-scorched earth, now became a moist and luscious oasis. The honey and butter had worked its wonders, elevating the flavor to new heights and turning the cornbread into a delectable treat that tantalized my taste buds.

This newfound trick, this simple act of drizzling honey, had granted me the cornbread of my dreams. No longer would I endure a dry and lackluster experience. Instead, I reveled in the pleasure of savoring beautifully moist, flavorful cornbread with every bite. The prospect of relishing the perfect tasting cornbread, rich with honey-kissed goodness, was no longer a mere aspiration but a delightful reality within my reach whenever the craving struck.

Now, armed with this culinary revelation, I can confidently enjoy the bliss of the perfect cornbread, elevating my dining experiences to new heights of satisfaction.

Chapter 2: Pressure Point

The year was 1970, a time when I was still a youthful figure scaling tall steel towers against a rugged landscape, the wild, open sky above me. On this particular day, the sun blazed relentlessly, casting long shadows across the construction site where a new power line was being erected across the state. I was part of the dedicated crew assigned to this monumental task.

Among us was Ron, the grizzled senior climber with years of experience etched into his weather-beaten face. With a nod of his head, he commanded my attention, "Joe, get the load line off." Without hesitation, I climbed over to the attachment point, wrapped my leg around the steel lattice face, and my muscles strained as I swiftly removed the chain. Little did I know; this routine action would trigger a chain of events that would leave an indelible mark on my memory.

In an unexpected twist of fate, the dozer on the ground unexpectedly backed up, causing a resounding crack to pierce the air. In that heart-stopping moment, an old power pole adjacent to our tower snapped at its base, hurtling towards me with terrifying velocity. I felt the impact as it struck me, pinning me against the angle iron and dragging me mercilessly down the steel tower. The pain was excruciating, my body suspended in mid-air, held hostage by the unforgiving forces of gravity.

Desperation flooded through me when the pole stopped moving and as I clung onto the steel lattice. My grip was the only thing preventing me from being torn in half. The fear was palpable, each breath a struggle to fill my lungs with air. There was no room left to maneuver, no escape from the imminent danger that loomed. In those terrifying seconds, I braced myself for the worst, prepared to face the unthinkable.

But miraculously, salvation arrived in the form of my courageous co-workers. Three of them sprinted across the precarious steel beams, their determination unwavering. With sheer strength and unwavering resolve, they pushed against the weighty power pole, just enough for me to slide out from its merciless grip. My torn shirt bore witness to the harrowing ordeal, but miraculously, I escaped with only a few scrapes and bruises. The power pole had come to rest on an eight-inch upright beam, a mere hair's breadth away from inflicting life-altering injuries. Three or four inches more, and it would have claimed my life.

Rough and tough from an early age, I dusted myself off, donned a spare work shirt, and returned to the task at hand. A couple of band-aids served as a reminder of the close call I had experienced. That evening, as exhaustion settled over me, I retired to bed a little earlier than usual.

Around 2 a.m., I was jolted awake by a dream that defied comprehension. It was as if my subconscious had woven together fragments of the day's events, morphing them into a puzzling narrative. I found myself in a hospital room, the sterile environment contrasting sharply with the reality of my motel room. Yet, the clock on the wall continued its steady ticking, its rhythm a disconcerting reminder of the inexplicable.

Lying there, in the dimly lit room, the ticking clock seemed to whisper secrets I couldn't yet grasp. It was a harbinger of something beyond my current understanding, a sign that the day's brush with danger had left an indelible imprint on my psyche. With a lingering sense of unease, I drifted back to sleep, and the mysteries of that strange dream remained unresolved.

Little did I know that this brush with mortality would shape my perspective in unforeseen ways, forever altering the course of my life.

Chapter 3: Fire in The Hole

In 2013, we were stationed in Battle Creek, Michigan, the southeast quadrant of the state, fittingly called the cereal city of the world by industry standards. Two cereal powerhouse companies forge the industry, making breakfast a convenient meal. And from its conception in 1906 to this very day, the remnants are still a hive of hustle and bustle, a breakfast staple that's here to stay. Yet in the heart of Michigan's southeast side, where the pulse of I-94 coursed through this community, east to west, or west to east, whatever the weary traveler's destination might be. It's the main-stay route across the southern end of the state, but for me it was "a road pounding four days a week commute." It was a sixty-mile drive each way, and this job would be for around five months, and for me it had become a run -of-the-mill scenario. The weather can also be a factor while commuting, and can change quickly. But overall I felt fortunate with this placement. Sometimes we would be on-call, or we would also travel elsewhere when storms knocked out power in other parts of the state.

My crew, which consisted of two co-workers and me, faced the perils of our trade, embracing the danger that danced on the edge of our daily lives. Four days a week, we toiled tirelessly, working trouble orders and braving the unpredictable nature of the job. Our safety protocols, etched in stone, stood as unwavering sentinels guarding our well-being.

Day after day, we strived to execute our tasks with precision, committed to delivering excellence to the utility company we served. As a contractor crew with the backing of a supportive union, we forged ahead with a shared purpose—performing our work safely and yet quickly. It was one muggy stormy night, and I remember it well. It was amid the tempestuous elements that I received a call from dispatch reporting an outage triggered by a fearsome lightning strike.

Navigating the winding roads by GPS, we arrived at the designated address, greeted by the fading presence of the fire department. Their valiant efforts were completed, having extinguished the fire with conductive water that raged through the smokey field. With their departure, we assumed control of the situation, left to confront the aftermath that crackled ominously before us. In the bright glow of our spotlights, we saw the troubled culprit, a seemingly harmless wire lay sprawled on the ground, a grotesque relic of its former self—melted, burnt into fragments resembling shards of an unholy mosaic. Yet the two other wires atop the pole hummed with electrifying energy, demanding our utmost caution.

The Lieutenant came up to me, and I said, "Good evening." He replied, “Yes, sir, will you be needing us for anything else?” The task at hand seemed straightforward. Splice the damaged wire with a specialized sleeve and re-energize it, and I confidently stated, “No sir, we’ll get it from here.” Larry, my diligent apprentice, prepared to venture into the field to retrieve the remnants of the once-vibrant conductor. Yet, as he took his first steps, I couldn't suppress the instinctual urge to ensure his safety. "Get your safety shoes on!" I urgently called out. It was my responsibility to perpetuate every job with undue caution, and Larry heeded my warning and swiftly retreated to the truck.

As I advanced towards the field wearing all my safety gear, I had the new springy wire in-hand, and treaded cautiously over the remnants of its melted predecessor. But then, as if the storm itself conspired against us, the high voltage unleashed its fury! The huge fiery snake broke the calm and quiet as the air was shackled by the crack of an explosive surge, and roared in vengeance from the 4800 volts that went straight to ground. In that heart-stopping instant, I relinquished my grip on the twenty-foot-long fireball that stubborn monster refused to extinguish. Fear coursed through my veins as I sprinted towards the truck, my hands finding solace in the insulated long-reach cutter. Inside the bucket and in unison and with a swift, calculated motion, we cut the dangling wire and brought an end to the blaze, which was harnessed by the power of my safety gear to conquer the fury that threatened to consume me.

In the aftermath of that harrowing moment though, and as the adrenaline began to ebb, I reflected on the gravity of the situation.

It was a chilling reminder of the thin line we walked between life and death, where adherence to safety measures could mean the difference between survival and catastrophe. Time itself seemed to freeze in those fleeting seconds, etching an indelible mark on my soul—a reminder of the delicate balance we maintained, the fragile dance we performed amidst the storm. Larry then asked,

"What do you think caused the wire to come down?" "It was lightning," I boldly told him, and also mentioned why. "A lightning bolt can carry around 300 million volts of electricity and about 30,000 amps of current," which was certainly a good explanation of the variables for the downed conductor. I then left that curious conversation at bay!

It was that sequence of events, and while we were in the bucket, I looked down and saw the culprit! There was a small limb on the bush next to the wire's dangling tail.

It was the perfect storm! The energized wire had an ever-so-slight air gap from the wet, live limb, and this wire's magnetic field was in its parameter of ignition. The new wire I dragged across the ground and gave this invisible magnetic field a means to ignite once again, and create a single circuit.

This invisible trigger operated! Pandora's box struck like the force of a speeding locomotive, the sound of a roaring volcano, and the quickness of a shooting star, and it all happened as fast as the blink of an eye.

This ordeal was just a part of how I thought throughout my life. But it was those flying saucers that Jim and I saw that almost daily I reflected upon mostly, and indeed, more clues surfaced as time rolled on, and I remembered or even surmised a little bit more about UFOs with each passing year.

Chapter 4: Tainted Water

In 2014, my path led me to a bustling city nestled in the northern and eastern reaches of Michigan, where the relentless pulse of I-75 reverberated through the concrete veins of its streets. Flint is also a fairly large city, with an interesting past. In reminiscing, I looked back to the 1800s, and in that time period, Flint was the leading manufacturer of horse-drawn carriages, earning Flint the moniker "vehicle city." Within this vibrant urban landscape, I continued my work as a contractor, tasked with handling trouble calls and shouldering the responsibility of being on-call at regular intervals. It seemed like a never-ending cycle of one call after another, with only brief reprieves snatched between jobs. Often, our hurried meals consisted of gas station fast food, a common choice being a quick slice of pizza and freshly brewed coffee, which was a suspect in essence. For me, coffee was almost a sparkling zinder that would pick me up, and gracefully let me down. But there was little to no exception to this formidable diet, and hidden in this honeycomb was another incident with major ramifications. It certainly seemed like nature had one of its surreal moments hidden in the walls of its fortress, and it was all about to come tumbling down!

As the summer days grew shorter, a dark cloud cast its shadow over my well-being. I remember that fateful day, etched into my memory like an indelible mark. The creeping onset of sickness plagued me daily, a persistent unwelcome visitor that wreaked havoc on my evenings. Concerned, I sought medical attention, leading to a seemingly never-ending series of events—surgery and emergency room visits—that cast a foreboding shadow over my future. Overhearing the doctor's whispered words, "he's not going to be getting better," sent a chill down my spine.

But I was not one to succumb easily. Dauntingly armed with a steely resolve, I defied the doctor's prognosis. The first two weeks of my battle were the toughest, with each day I had a formidable challenge. On the initial days, my food intake dwindled to meager portions of chicken noodle soup, a frail sustenance to fuel my weary body. As the days progressed though, I gradually introduced crackers to accompany my soup, a small glimmer of progress amidst the storm.

With unwavering determination, I persevered, gradually expanding my culinary repertoire. Day by day, I cautiously ventured into consuming small, full meals that eschewed grease and unhealthy fats. It felt like a step in the right direction, a flicker of hope in the midst of uncertainty. Yet, each passing month brought occasional bouts of discomfort, a reminder that the journey to recovery was fraught with obstacles. Nevertheless I pressed on, and my digestive system grappled with the demands placed upon it.

Yet, and to this day, I approach my eating habits with utmost care and consideration. I have become acutely aware of the delicate balance that governs my well-being. The grandeur of this achievement was my new diet, which consisted of coffee, water, and fruit drinks, while carbonated beverages are forever banished from my culinary repertoire. The experience has left an indelible mark, a constant reminder of the fragility of our health and the importance of nourishing our bodies with conscientious choices. I reiterate and sometimes think about, in hindsight, how civilizations have progressed! I think about how the horse and carriage were a luxury item for most people, and now travel is a necessity that most people can't live without. Flint was a catapult for travel, and now we're even looking at traveling to other planets. Science also lets us see deeper into space. We can now examine other planets with advanced photography, and even study the atmospheres of other worlds, and one can only wonder what's next in our journeys of life.

Chapter 5: Slip and fall

This year's winter, 2022–2023 was unusual, with temperatures oscillating between peaks and valleys, following an unwavering pattern. It was as if the hands of time toyed with the mercury, teasing us with moments of moderation and sudden spikes. The approach of the spring equinox on March 19th, 2023, brought a glimmer of hope, as temperatures moderated and the sun's rays grew more direct. The weather seemed to gravitate closer to the realm of normalcy, a stark contrast to the frozen chaos that reigned just two months prior. Yet, in the depths of my memory, it felt as though time itself had momentarily stood still.

On a bitterly cold night in March, Lizz arrived home after a shopping trip, her car laden with grocery bags. As Izzo, our faithful Shih Tzu companion, and I made our way to the back door, the darkness enveloped us, punctuated only by the gentle glow of the porch light. The back porch deck appeared relatively safe, its surface glistening slightly from earlier moisture.

Lizz emerged from the car, her hands full with bags, and I eagerly stepped onto the deck to lend a hand. Little did I know that treacherous black ice lay concealed beneath my feet. In an instant, the world around me spun and flipped, my body tumbling onto the deck with a resounding thud. My head collided with the unforgiving surface, while my left shoulder absorbed a direct blow. Slowly, I gathered my senses, gingerly shifting my body to a seated position. With Lizz's assistance, I managed to stand, relieved to find no pains associated with broken bones.

Wasting no time, I hurried to the garage, seeking a solution to prevent future accidents. Armed with a bag of rock salt, I diligently coated the deck, determined to make it safe. Lizz, a nurse by trade, kept a watchful eye on me that evening, ensuring my well-being.

The following morning, I awoke early, expecting to face the aftermath of the fall. To my astonishment, I felt fine. How could this be? I wondered, marveling at the resilience of the human body. Our daily routine resumed, as though nothing out of the ordinary had occurred.

Days turned into weeks, weeks into months, and our routines became entrenched in the fabric of our lives. However, as the mundanity of everyday existence settled in, my mind began to drift towards a couple of television shows I had recently watched—a series delving into the realms of science and history. These shows, with their discussions of alien abductions, struck a chord within me, stirring memories of my own encounter. The sightings of unidentified flying saucers suddenly flooded back into my consciousness.

Two particular episodes seized my attention as they explored the enigmatic world of alien abductions. Although the shows lacked clarity regarding the experiences of the abductees, they left me pondering the possibility that something extraordinary had indeed happened to me. Attempting to piece together the timeline, I recalled a bright white light, but crucial details remained elusive. Desperate to unravel the truth, I retraced our every movement, hoping for a revelation. Alas, my efforts proved fruitless. Frustration settled in, until a moment of realization struck me like a bolt of lightning. I thought about the flying saucers we saw and a hidden answer was revealed right before my very eyes!

My neck should have been sore after gazing upwards for thirty to forty-five minutes. Yet, there was no trace of discomfort. It became clear that something profound and terrifying had occurred, something beyond the reaches of my memory. As I grappled with these chilling facts, a different history show shed some light on this puzzle. The broadcaster's words echoed in my mind: "Some alien abductions occurred, and the people did not know it happened." A shiver ran down my spine as I realized the implications of this revelation.

With newfound certainty, I pondered the enigmatic encounter that had altered the course of my existence. That night, as the clock struck ten, I retired to bed, drifting off to sleep without a hint of apprehension. Yet, around 2 a.m., I awoke abruptly, compelled to visit the restroom. Disoriented and groggy, I shuffled back to my bed, longing for the solace of sleep. But just as I closed my eyes, a blinding flash of light shattered the tranquility of the room. Terror engulfed me as I found myself enclosed within a brilliant white space, floating weightlessly alongside Jim, my companion. Paralyzed, unable to move, we hovered in a tight, confined room, approximately four to five feet high and barely a foot off the floor. Five glowing figures materialized before us—aliens. With their immense blue eyes, slim bodies, and elongated arms, they stood at a mere three feet in height.

As the alien in the center held a glass container in one hand, a chilling silence descended upon the room. With the press of a button, a puff of white smoke emerged, slowly transforming into two equal, round clouds. These ethereal formations swiftly morphed into rope-like wisps, hurtling toward us with alarming speed. One headed for Jim, the other for me. The rope cloud transformed into a sinister shade of gray before vanishing into my nostrils.

Startled awake, drenched in sweat, I leaped out of bed, frantically flicking on the lights. Staring at my reflection in the mirror, I appeared unharmed, save for the weariness etched across my face. That was the last coherent memory I could grasp from that eerie night. Sleep eluded me, as a cascade of fragmented recollections flooded my mind. I recollected an incident after vaulting over a fence and sprinting into the woods—a faint stinging sensation in my arm that I had dismissed as mosquito bites. Now, it seemed that something far more unsettling had transpired, leaving me with the disconcerting notion that something had been taken from me, or something put into me!

Chapter 6: Identity Unraveled

Present day, life is the same for me. Although the pleasures of life don't make sense anymore. It's not the same when I first tried cornbread, or even the peak of my food addiction. Life was bleak with no purpose, except the nightmares with the blinking lights.

I found myself sitting at my kitchen table, staring at a plate of cornbread. It used to be a delight, warm and inviting. But today, it's just a lifeless piece of bread on my plate. I took a bite, chewed without savoring, and swallowed without tasting. It's as if my senses have grown dull, unable to appreciate the simple pleasures of life.

My phone buzzes, breaking the silence. It's a message from my friend Sarah, asking if I want to go out for a drink later. I considered it briefly and then declined. The idea of feigning enjoyment in social settings had lost its appeal. Life, once vibrant and full of purpose, has now become a bleak and colorless existence.

There was a time when no one could have stopped my restless personality from achieving anything; if there was a party, a meetup or even a friend in need of some company, I was always there to help out, even if I was tired and at my wit's end. Going out and having fun was comfort and solace for me. It brought me peace and made me feel exquisite and fun; a complete contrast from the person I've become today. Each day was a carbon copy of the last, and I felt like a ghost drifting through life.

As I sat there, lost in thought, I couldn't help but wonder where it had all gone wrong. One night, while lying in bed, unable to sleep, I made a forlorn decision. This empty existence cannot continue. I needed to find meaning, to understand where it all went wrong. I reach for my laptop, determined to retrace my steps and unearth the clues to my current despair.

I sifted through my memories, trying to pinpoint the moment when everything changed. It hits me like a sudden jolt of electricity – the night with the blinking lights. I ventured into the unknown, hoping to discover something extraordinary, and now I can't even remember where I went that night.

Hours turned into a blur as I scoured my browser history, searching through photos, and delving into my old journals. But there's nothing! It's as if that night has been erased from existence. Frustration and desperation washed over me, and I realized that I must uncover the truth of that fateful night, the key to my desolation.

Weeks blended into months, and my obsession with unraveling the mystery intensified. I've become a recluse, consumed by my quest for answers. Friends and family expressed their concern, but I couldn't explain the driving force behind my relentless pursuit.

One evening, my phone rang, and it was Sarah, my estranged friend from before, pleading with me to come out, to let her help me break free from this self-imposed isolation. I was torn between my relentless quest and the yearning for human connection. She was one of my favorite buddies, and I do miss her.

But reluctantly, I agreed to meet her at a local coffee shop. The familiar scent of brewed coffee and the cozy ambiance welcomed me as I entered. Sarah was waiting at a corner table, her eyes a mix of relief and worry.

"It's been too long," she said, embracing me tightly.

As we sat down, she gently probed, "What's been happening? You've withdrawn from everyone, lost in your own world. We're all worried about you."

I sighed, and had a weary look in my eye. It's very clear at this point to everyone, including Sarah that I'm not okay. "It's okay, you can let me know that I'm always here for you," she caringly said while coming closer to me and caressing my hand. Sometimes, I felt that she might have had a crush on me, but mostly I think this is just my mind playing tricks on me again.

But since her eyes remained concerned, I finally made the effort to open up about my search for answers. Taking a deep breath, I start from the beginning. I told her about the blinking lights, the white space, and the void that had engulfed my life. Sarah listened to it all attentively; her concern was very evident, and after I was finished, she took a minute or two to absorb my story... and it was then that I thought she was thinking I'm delusional too.

"I can't promise to help you find the answers you're looking for," she finally blurted out, "but together, we can find meaning in the present. Let's rediscover life's pleasures, even if they don't make sense right now."

I listened to her intently, but it just felt like she did not understand me or trust my words, just like the rest. "Okay, I promise," I replied, just for the sake of responding. After this meeting, something did change in me. I wanted to prove myself right. I did see the blinking white lights, it was true. Right? I had to find out where it was, and where they took me that night.

The next day after meeting Sarah I blocked her on my number and went back to my search. This time, I waited till late night and then started to walk outside into the darkness. I was going everywhere I knew in search of a white light. 'Where was it?' I whispered to myself. Unfortunately, that day went into a complete vain of emptiness.

Upset and frustrated, I decided to visit a friend's house, Jim, hoping that he would be able to help me out as I recall him being there too. I rang his doorbell, and when the door opened, I'm greeted by a puzzled expression.

"Can I help you?" my friend, Jim, asked cautiously.

"It's me, Joe," I say, taken back by his lack of recognition.

Jim furrowed his brow. "I'm sorry, but I don't know who you are."

A chill ran down my spine. This isn't the friend I remember. A growing sense of fear and confusion set in.

“Um, maybe I’m at the wrong house. Sorry.” I forsakenly told him, while having no idea why Jim wouldn’t recognize me. I did turn to leave, but hesitation came over me. “Are you sure?” I asked again.

“Sure of what?” Jim responded, a bit defensive this time, like he was being threatened and felt unsafe in my presence. It was so unlike him.

“You don’t know me, correct?”

“No, I don’t. Now please leave before I call the police!” Jim said sternly and shut the door. His action took me back and my anxiety continued to build. I could feel it in my bones; everything was shivering, and my brain was confused. The world around me seemed to shift, and my perception became distorted. I couldn’t shake the feeling that something was terribly wrong.

Trying to understand Jim's perspective, I reluctantly returned home, disheartened, and decided to take a long shower, hoping it would clear my mind. As the water cascaded over me, I began to feel a profound unease. It's as if my very identity was slipping away.

After the shower, I caught a glimpse of my reflection in the mirror. My heart pounded in my chest as I stared at the face before me. It was not mine. The face looking back at me was unrecognizable, a stranger's visage.

I stumbled back in shock; my mind raced with fear and confusion. Nothing made sense anymore. The nightmares with the blinking lights, the loss of identity, the transformation – it was a bewildering puzzle with no solution in sight. "Who are you?" I said, while touching my face. This was not my face, my nose, my lips, eyebrows, or even hair! Who was this? Who am I?

Chapter 7: Inexplicable Clues

I left the bathroom in a daze, my mind spinning with confusion and fear. I had to find out what was happening to me, to understand the inexplicable changes that had taken place in my life. Walking through my small apartment, I heard a sudden, jarring noise coming from the closet. It was a sound that didn't belong in my quiet living space. Cautiously, I approached the closet and swung the door open. My eyes widened in shock at the sight before me.

Inside, I found a knife, its blade glistening ominously in the dim light. Beside it lay a bloodied napkin, as if it had been hastily discarded. My hands trembled as I picked up the napkin. It was stained with a deep, dark red liquid, and as I unfolded the napkin, my heart sank. The word “garage” was written on it! Was it a clue, or perhaps a warning? I couldn't be sure, but I knew I had to follow it. My quest for answers had taken an even more interesting turn.

With a sense of urgency, I rushed to my building's underground garage. But as I entered, I was met with a sight that left me utterly bewildered. The garage was not the one I remembered; it belonged to someone else entirely. It was sleek, sterile, and filled with cars that I could only describe as futuristic.

"Excuse me," I stammered, approaching a person who appeared to be a mechanic. "I think I'm in the wrong place. This doesn't look like my garage."

The mechanic turned to me, his face was a mix of curiosity and suspicion. "Wrong place? Are you lost?"

I nodded, my head swimming with confusion. "Yes, I must be. I don't recognize any of this place."

The mechanic's gaze bore into me for a moment before he finally spoke. "You must be in a coma or something. This is how things have always been in Coverville."

“Cover..what?” I could not believe this. This was not the town I just was in a few minutes ago.

Then I saw a calendar hanging behind the mechanic, very conveniently telling me the year.

“2..0..5..0. 2050?” I said in disbelief. The mechanic was looking at me weirdly now. His eyebrows arched up in a position that suggested he thought I needed help.

I stood in the futuristic garage, utterly stunned by the mechanic's revelation. "Things have changed a lot, buddy. We're in the year 2050," he said, trying to help me make sense of it all.

The words hung in the air, and I struggled to comprehend the enormity of what he was saying. It couldn't be true, but as I stepped outside my garage, my world had transformed into something beyond my wildest imagination.

Gleaming skyscrapers towered above me, their glass surfaces reflecting the shimmering lights of flying cars that soared overhead. Robotic humans moved about with mechanical precision, and the streets were filled with technological marvels that defied explanation. The entire scene felt like I had been transported into a futuristic sci-fi television show.

As I tried to comprehend the unbelievable, my mind raced back to the napkin and the knife in my closet. What has led me here? What caused my face to change so dramatically? And who had left that ominous message on the napkin?

In search of answers, I decided to visit a nearby restaurant. The exterior was sleek and automated, with no visible staff. A holographic menu hung in the air, and I cautiously approached it.

"Welcome to FutureFlavors," a disembodied voice greeted me. "Please make your selection."

I hesitated but eventually made my choice. "I'll have the classic burger, please."

The voice acknowledged my order, and I watched in bewilderment as a 3D printer nearby sprang to life. Layers of synthetic ingredients materialized, forming a burger that appeared as real as any I had ever seen. It was a surreal experience, one that left me feeling disoriented and disconnected.

When the burger was ready, a robotic arm placed it on a tray, and I paid somehow with probably facial recognition at a self-service kiosk. As I took a hesitant bite of the burger, my frustration and anger grew. The taste was all wrong, it was a pale imitation of the real thing.

"This is absurd!" I muttered to myself, my voice filled with frustration. I couldn't bear the thought of this bizarre future any longer.

I left the restaurant, my heart pounding with a mix of emotions. As I returned home, I decided to try and sleep, hoping that it would provide some respite from the overwhelming strangeness of the world around me.

I laid down in my bed, closed my eyes, and hoped that sleep would offer some solace, some semblance of normality. As I drifted into unconsciousness, I whispered to myself a comforting self-assurance.

“Maybe this will fix everything.”

Chapter 8: The Unending Nightmare

The following morning, I awoke in my familiar bed, sunlight streaming through the curtains. For a fleeting moment, I believed that the strange events of the past days had all been a dream, a bizarre hallucination born of stress or exhaustion. I sighed with relief, thinking that I could return to my ordinary life.

But as I got out of bed and went about my routine, a sense of unease lingered. The world outside appeared to have returned to its mundane state, with no flying cars or robotic beings in sight. The taste of the synthetic burger, however, still haunted my memory, and doubts persisted.

I decided to check my reflection in the mirror, half-expecting to see the face of a stranger once again. To my surprise, my reflection appeared as it should be, my own face staring back at me. The relief was palpable, but a nagging uncertainty remained.

As the days passed, I tried to convince myself that I had indeed been caught in a bizarre dream, a temporary delusion that had now dissipated. Life seemed to regain its familiar rhythm, and I clung to the hope that everything would return to normal.

But gradually, an unsettling realization began to dawn on me. The dreamlike reality I had experienced may not have been a mere aberration. It felt all too real, and the world around me showed no sign of reverting to what I had known. Doubt and confusion gnawed at my sanity, and I struggled to determine the truth.

Desperation and despair intensified with each passing day. The line between reality and illusion blurred, leaving me trapped in a nightmarish existence from which there seemed to be no escape. It was an absurd thing to say, but as weeks passed on, and strangely enough, I wanted to go back to the place. It was something that allowed me to feel almost excitingly different.

Chapter 9: A New World

Never had I thought I would reach a state where the routine of daily life became my greatest enemy. Each day blended into the next in a monotonous cycle of waking up, eating the same food, heading to work out of necessity, and returning home to the solitude that had become my unwelcome companion. It was the sameness of it all that wore me down: a relentless grind of predictable weekdays and a fleeting escape into television and more food on weekends. I hated the unending sameness, the suffocating ordinariness of it all.

But deep down, hidden beneath my resistance, a yearning persisted—a longing for the inexplicable, the otherworldly, the feeling of exhilaration I had experienced that one unforgettable night. I wanted to go back, even though it defied reason and seemed utterly absurd. That night had taken me to a place that allowed me to feel different, alive in a way I hadn't thought possible.

One ordinary evening, as I lay down expecting to wake up to yet another mundane Monday, my world was once again disrupted. This time, it wasn't a strange taste or fleeting memory; it was a bright, almost blinding light that flooded my vision. I blinked against the brilliance, struggling to comprehend what was happening. As my blurry vision gradually cleared, an inexplicable portal or gateway came into focus.

The portal was unlike anything I could have imagined. It shimmered with an eerie magnetic energy, shifting and undulating as if it were a living thing. It moved, not in a straight line, but in a trippy, maze-like pattern, as if beckoning me to step through its enigmatic depths.

With a mixture of trepidation and longing, I took a deep breath and walked towards the mesmerizing portal. Each step felt like an eternity, the air around me humming with an otherworldly energy. I hesitated for a moment, looking back at the world I had grown weary of, and then, with a resolve born of curiosity and yearning, I stepped through the force-field of energy's opening.

The sensation was disorienting, like being caught in a whirlwind of colors and emotions. When I emerged on the other side, I found myself in a world that was both new and hauntingly familiar. It was the same world I had been craving—the world where my whole body seemed to come alive.

Every detail of this place was vibrant and teeming with life. The sky above me was a mesmerizing blend of hues I couldn't name, and the landscape stretched out before me in a breathtaking panorama of colors and shapes that defied earthly logic. The air tasted different, charged with a vitality that invigorated every cell in my body.

As I ventured further into this strange yet captivating world, my personality seemed to shift and adapt. The weight of my mundane existence lifted, and I felt a newfound sense of freedom. With a spontaneous burst of energy, I found myself drawn to a nearby pub, its neon sign beckoning me like a siren's call.

Inside, the atmosphere was electric. Conversations in languages I didn't understand buzzed around me, and the vibrant blend of music and laughter filled the air. The bartender greeted me with a warm smile and a curious look, as if recognizing a newcomer to their world.

I took a seat at the bar, ordered a drink I couldn't pronounce, and allowed the sensations of this extraordinary place to wash over me. For the first time in what felt like an eternity, I felt truly alive, and caught in that moment of exhilaration and uncertainty, but most of all, a sense of belonging in this new world.

As I sipped my drink and lost myself in the kaleidoscope of experiences around me, I couldn't help but wonder if I had found my true home, a place where the line between reality and illusion was beautifully blurred, and where I could finally escape the confines of the mundane existence I had grown to despise.

As I sat at the bar, immersed in the vibrant atmosphere of this new world, my attention was drawn to a woman unlike any I had ever seen before. She moved with an effortless grace, her long, ebony hair cascading down her back like a waterfall of midnight silk. Her eyes were a shade of deep emerald green, framed by lashes that seemed to glisten in the ambient light of the pub. Her beauty was both ethereal and magnetic, a presence that demanded attention.

She noticed my gaze and turned to meet my eyes with a knowing smile, as if she had been expecting me. Her voice, when she spoke, was rich and entrancing.

"You must be new here," she said, her words laced with a hint of intrigue.

I nodded, still captivated by her presence. "Yes, I just arrived. It's... it's unlike anything I've ever experienced." I

She chuckled softly, a sound that sent shivers down my spine. "That's the allure of this place. It has a way of drawing you in, doesn't it?"

I couldn't help but be drawn further into the conversation with her, and my curiosity grew with each passing moment. "You seem to know your way around here. Do you come here often?" I asked.

She nodded, and her eyes held a glimmer of mystery. "I've been here for quite some time, and I've seen many like you come and go."

The night unfolded with stories and laughter, which the two of us shared tales of our respective worlds and the adventures we had experienced. It was as if we had known each other for a lifetime. Our connection deepened with every word exchanged.

As the evening wore on, I found myself unable to resist the temptation any longer. "Would you like to go out with me sometime?" I asked, my heart pounding with anticipation.

She smiled, a hint of mischief dancing in her emerald eyes. "I thought you'd never ask."

We left the pub hand in hand, the vibrant streets of this new world stretching out before us like a playground of endless possibilities. The night was alive with energy, and we explored it together, savoring each moment as if it were a precious gem.

When we finally arrived back at my place, the connection between us was undeniable. We didn't need words to convey what we both felt. As we lay together, the world outside seemed to fade into insignificance, and in the morning, I was still in this mesmerizing new world.

“What’s your name?” I asked the woman while she got dressed to leave.

“Claire,” she softly replied while smiling. A smile that grew warmth inside me. It was beautiful.

“Would you meet me again?” I asked, without even thinking. How stupid of me! I thought.

“Of course Joe!”

I smiled back at her as she kissed me goodbye and left.

"That's weird. I don't remember ever telling her my name," I thought.

Still, I couldn't have been happier. The desire to see her again fueled my excitement, and our bond continued to deepen with each passing day. We went on more dates, exploring the wonders of this place together, whether it was dining at exotic restaurants, dancing under the dazzling stars, or simply strolling through its enchanting streets.

As our connection grew, so did my reluctance to return to my old life. The mundane existence I had once known lost its appeal entirely. There was a magic to this new world, an irresistible allure that kept me coming back for more. The taste of the synthetic burger and the memories of my former life had faded into obscurity, replaced by the enchantment of the woman by my side and the enchanting world that had become my new reality.

With each passing day, I became more entwined in this captivating new world, and it was a journey I was eager to continue, for I had found something here that I had longed for all my life—a profound connection with the woman who had become my muse, my confidant, and my greatest adventure.

Chapter 10: The Vanishing of the Old World

Five years had passed since I had last glimpsed the other world—the one that felt like an impossibly vivid dream. In my current reality, the idea of manually making my bed or brushing my teeth seemed like a distant memory. Everything was automatic here, including chores and daily activities. The comfort was anything like the other world. This really is the ultimate dream of anyone alive. Now, my wife Claire and I have also embraced the conveniences of our time, and Tenisha, our trusty household robot, ensured that our daily routines ran seamlessly.

It is surprising to even share that I'm married now. The proposal was simple. A year after our dating life began, I knelt down to ask Claire's hand in marriage. Luckily, she did not have a proper family, so we did not have to go through the boring traditions and official obligations. Within a week, we were married, and by the next one, we had our daughter, Kiara. Timothy was a surprise though; we weren't planning a new member this soon, but I guess who could've complained when such blessings were on the way. Our family had grown, and I couldn't help but feel a profound sense of fulfillment that had eluded me in my previous life. Claire and I had initially only expected one child, but life had surprised us with a second.

Now, our home was filled with the laughter and joy of our two children, Carl and Kiara.

Waking up to the sound of their laughter, I would greet them with a warm smile. "Good morning, my little adventurers," I'd say, ruffling their hair as they bounded into our bedroom. Their eyes sparkled with excitement, and I relished every moment of their boundless energy.

Claire, the love of my life and the rock of our family would already be in the kitchen, whipping up a delicious breakfast. "Pancakes or waffles today?" she'd ask, her eyes twinkling mischievously. Our children would chorus their preferences, and we'd share a hearty family breakfast, filling our home with the aroma of freshly cooked food and the warmth of togetherness.

After breakfast, I'd drop Carl and Kiara off at school. As I watched them disappear into the schoolyard, I couldn't help but marvel at the life I now had. It was a life of routine, of predictability, but it was also a life filled with love, support, and the kind of fulfillment I had never imagined.

On most days, I found it easy to dismiss the memories of the other world as nothing more than an extravagant dream. The simple joys of coming home to the laughter of children, the unwavering support of Claire, and the bonds we had forged made it all seem distant and inconsequential. But then, one fateful night, as I kissed Claire goodnight and settled into bed, something inexplicable happened. Without my consent or understanding, I found myself back in the portal. When I awoke in the morning, I was back in the old world, the world I had come to despise.

Fury coursed through me as I screamed in frustration. I wanted my wife, my children, the life I had grown to cherish. I rushed outside, only to find that the old world had changed beyond recognition. It was the same streets, yet not the same. The buildings were dilapidated, and an eerie silence hung in the air.

Checking my phone, I discovered a flood of emails and missed calls, most of them from family members. My heart raced as I read the messages. My parents had passed away, and everyone I had known had vanished without a trace. It had been more than five years since my disappearance. Maybe even twenty...I had no clue anymore.

In disbelief, I made my way to the graveyard, my heart heavy with dread. There, I found the graves of my parents, my siblings, my friends, and even their children. The headstones bore witness to the passage of time, and I was overwhelmed by a grave sense of loss and grief. I stood before my mother and father's grave. Tears welled in my eyes. "I'm so sorry I wasn't there," I whispered, my voice breaking. "I never got to say goodbye." The graveyard was a silent testament to the passage of time, a stark reminder of the life I had left behind, and the loved ones I had lost.

I knelt before their graves, the weight of my absence crushing me. "I don't know how or why this happened," I murmured, tears streaming down my face. "But I promise, I will never forget you. I will carry your memory with me always."

The weight of my apparent immortality bore down on me like an unrelenting burden. I watched as the seasons changed, the years passed, and the world continued to evolve without me. It was a tormenting paradox—I was trapped in a state of eternal existence while the world around me aged and decayed.

Depression settled in like an unwelcome guest, its grip tightening with each passing day. The memories of my family, my wife Claire, and our children, Carl and Kiara, haunted my every waking moment. I longed to hold them, to hear their laughter, to feel their presence, but they were forever beyond my reach.

The isolation gnawed at my sanity, and I found myself grappling with questions that had no answers. Why had I been brought back to this desolate world? What purpose did my seemingly eternal existence serve? The absence of meaning weighed on me, a heavy shroud that smothered any flicker of hope.

Days turned into weeks, weeks into months, and months into years. Time lost its meaning in this desolate existence, and I felt like a ghost, condemned to wander a world that had long forgotten me.

The depression deepened, its tendrils snaking through my thoughts, poisoning my soul with despair.

In my darkest moments, I yearned for an end to the torment, for a way to escape the endless cycle of isolation and confusion. But as I searched for answers, and found none. I was trapped in a purgatory of my own making, forever separated from the life I had cherished and condemned to grapple with the immortality that had become my curse.

Chapter 11: Descent into Darkness

I knew what I had to do now, and what needed to be done with my body: burn it off the surface of earth. How else would I go back to Claire? Or be able to disappear from this desolate world. I don't want to live anymore. I thought to myself. It had been so long since I woke up with Claire beside me, I tried multiple times to achieve the result. On most days now, all I did was sleep in my bed to make sure I woke up in the world I truly belonged to and was myself. My world. Overwhelmed by the relentless depression and suffocating isolation, I found myself standing on the precipice of despair.

The memories of Claire, Carl, and Kiara, once a source of warmth and solace, had become a painful reminder of what I could never have again. The thought of returning to my old life seemed impossible, and I couldn't bear the weight of my apparent immortality any longer. I had exhausted every avenue of hope, leaving only one grim option in the shadowy corners of my mind. Suicide beckoned like a siren, promising release from the torment that had become my existence. The idea of ending it all became a haunting obsession, a way out of this seemingly endless nightmare.

One fateful night, consumed by despair and the conviction that there was no way back to Claire and our children, I took that desperate plunge into darkness. With trembling hands, I carried out my final act, expecting it to bring an end to my suffering. I went to the nearest store and asked for a rope. A simple but strong rope, paid for it with the cash that was left and came back home. It wasn't as easy as the films make it out to be; there is no fulfillment while doing it, only dread and despair. I thought it would be just one click of a rope, but no, as soon as my head jerked up and I threw the chair underneath me, my head went blank. No struggle, no confusion, no pain, just blank.

Instead of the anticipated oblivion, I awoke on an alien table, surrounded by strange beings bathed in an eerie, blinding white light. Panic surged through me as I struggled to make sense of my surroundings. A surreal and unsettling reality had replaced the pain and confusion that followed my suicide attempt.

The beings before me were unlike anything I had ever seen, their forms both mesmerizing and unsettling. Their eyes, if one could call them that, held a depth of knowledge and intelligence that transcended human comprehension.

One of them, with elongated fingers and iridescent skin, stepped forward and spoke in a language that seemed to bypass my ears and echo directly in my mind. "Welcome, traveler," they said, their voice a symphony of alien tones. "You have embarked on a journey few have ever undertaken."

I stammered, struggling to find my voice. "Who are you? Where am I?"

The being extended a limb toward my head, and I flinched, but instead of pain, I felt a strange sensation within my skull. It was as though something had been implanted there, something that connected me to these enigmatic beings.

"We are the Selenari," the being explained. "And you, dear traveler, have been chosen for a unique purpose. We have implanted within you a device that bridges the gap between our worlds. You possess knowledge and experiences that are invaluable to us."

My mind reeled as I tried to process this revelation. The suicide that I had believed would be my escape had instead led me to a destiny I could never have foreseen. I was a conduit between two worlds, a living bridge between the familiar and the incomprehensible.

The only words I could truly muster up were, “What in the world is going on?”

Made in the USA
Coppell, TX
05 November 2026

11331607R00038